CLIMATE CHANGE ENIGMA
A DELICATE BALANCE

Dr. Abdulla Bilhaif Alnuaimi

GULF BOOK
SERVICES

Published by **Gulf Book Services Ltd**
20-22 Wenlock Road, London,NI 7GU,UK
Email: info@gulfbooks.co.uk
Office No: G23, Sharjah Publishing City Freezone
Sharjah – UAE

GULF BOOK SERVICES

First Published in Gulf Book Services Ltd
Illustrations © composed by Madhavi.S

ISBN: 978-1-7384322-6-4
Year: October 2024

Typeset in Garamond by Forging Minds, India

Contents

Author's Note

My parents and elders often narrated stories about their carefree lives in a clean environment – gazing at the blue sky during the day and admiring the glittering stars at night. They never spoke of 'environmental degradation' because of burning fossil fuels or 'smog' lasting for hours in bright sunshine.

These alarming notes came into our vocabulary after nations trained their economic growth on the Industrial Revolution and the concrete jungle.

The situation stands alarming as we read, hear, or watch the unyielding protests by climate change activists – with some climbing buildings, bridges, and even famous monuments – to make their voices heard by their respective governments.

The clarion call of 'Save The Planet' is not an individual crusade; but a collective decision of our generation and beyond to safeguard the environment.

It's time to learn from nations fascinated by bicycles: the young and the old pedal to schools, offices, supermarkets, parks, and scenic countryside.

Imagine, if we all cycled, the global carbon emissions would drop by nearly 700 million tonnes annually! While sounding constructive and interesting, it offers twin benefits – health and joyousness.

But, as of now, the world stands at a critical juncture. It faces the most intricate and pressing challenges on climate change that we could have ever imagined.

The 'smoke canopy' covering the globe strikes at the core of the planet's equilibrium, affecting the energy system, well-being, and the rich tapestry of biodiversity in forests and oceans.

The quest for cleaner and greener energy extends beyond alternative energy sources. Though the demand for cleaner air is holistic; churning out environment-friendly energy is technically challenging and expensive.

In the following chapters, the causes of Earth's energy imbalance are discussed, vis-a-vis what should be done for the next generation to breathe in a better environment.

The industrial growth, fuelled by fossil fuels across the globe, led to the rise in carbon emissions that brought the climate change debate from the United Nations and activists to our dining table.

The intensity of hurricanes and the consequences of floods are scary but seldom leave a lasting impression on us. We raise eyebrows in shock on reading the statistics, but how many ponder why heavy rainfall inundated the UAE during spring, the Arctic Circle icebergs melt during summers, and pollution levels rise in New Delhi during winters?

Unending questions and lengthy debates between cheap fossil fuels and costly renewable energies are here to stay unless an alarm is raised that the Earth is heating up, and soon the planet may have to be put on a 'gasmask'.

Children should be educated on the disequilibrium motorised by the escalation of greenhouse gases (GHGs) that ensnare the heat in our atmosphere, leading to warmer climates throughout the year.

The book, therefore, sheds light on the repercussions of climate change on our environment and societal structure. The thawing of glaciers to the rising sea levels – is an advance notice of the amplifying of catastrophic weather phenomena that could lead to the proliferation of life-threatening diseases, floods, droughts, and scarcity of crops by 2050.

I have tried to provide an insight into the emerging menace of novel gases and their potential threat to warming trends. It is conceivable that synthetic gases may soon eclipse today's dominant greenhouse culprits — carbon dioxide (CO_2) and methane (CH_4). Also, nitrous oxide (N_2O) looms as a possible formidable adversary in the Earth's climatic saga.

Swift and collective steps are imperative to curtail GHGs to adapt to the evolving climate and fortify societal resilience.

My aim was to simplify the understanding of the impact of climate change, and I hope making a complex topic easy to grasp will enlighten the readers.

The text's premise is to debate existing and prospective solutions while confronting the limitations of current strategies, such as the finite nature of renewable resources, complexities of green finance, and intricacies of carbon markets.

Fostering an inclusive dialogue, the book recognises the diverse viewpoints, hurdles, and prospects accompanying various approaches in harmonising the future energy landscape. Readers are invited to engage, suggest, and offer insights on these collaborative scholarly pursuits.

Let's begin from Me to We...

Dr. Abdulla Bilhaif Alnuaimi
Sharjah, United Arab Emirates, 2024

1

The Deep-Rooted Muddle

Climate change is real. The science is compelling. And the longer we wait, the harder the problem will be to solve.

John F. Kerry,
former United States Secretary of State

It is said that starting strong is good, but it is epic if the mission is accomplished.

Following the dictum, on a highly positive note, in 2009, at the 15ᵗʰ Conference of Parties (COP15) of the United Nations Framework Convention on Climate Change (UNFCCC) in Copenhagen, Denmark, explicit emission pledges were deliberated by all major economies. The accord saw the adoption of a new set of international goals.

The Kunming-Montreal Global Biodiversity Framework (GBF) was agreed upon by 188 governments, who committed to address the ongoing loss of terrestrial and marine biodiversity.

Later, the Paris Agreement of 2015 became a landmark achievement in the global fight against climate change, as 197 countries consented to limit global warming to below 2°Celsius (C) or preferably 1.5°C, compared to the pre-industrial (1850-1900) levels.

Importantly, the developed nations were willing to achieve a collective goal of mobilising $100 billion annually by 2020 and lasting through 2025 for climate action in developing countries. However, during the subsequent COP meetings, the push and pull factors on the finance front became evident between the rich and the poor nations.

Doubt Separates. Trust Unites

Though, in 2023, the Organization for Economic Co-operation and Development (OECD) stated that $100 billion was likely delivered in 2022, many developing nations and their consortia maintained they were yet to see the money. They added that most of the capital came to them as loans, not as grants, increasing their debt burden.

The countries questioned the promises made by the

wealthy nations and criticised them for defaulting on their word.

While dealing with this scenario at COP28 in Dubai, a historic agreement was reached to start funds for addressing loss and damage associated with the adverse effects of climate change. It was decided that the Green Climate Fund would be parked in the World Bank for a pre-determined interim period, following which it would become an independent entity.

The world leaders assured that the World Bank would transform and made tall promises to mitigate the climate crisis, but the ground reality once again showed that the rhetoric was a disaster.

According to the Third World Network, an independent non-profit international research and advocacy organisation, involved in issues related to development, developing countries, and North-South affairs, the rich versus emerging nations feud is about who should contribute to the new quantified goal on climate finance.

In some countries, the international response was muted as war, inflation, and poor governance drove the poorest to the brink of famine. Experts are rightfully questioning that if the world cannot offer food to starving children, it is unlikely they will unite to defeat climate change.

The UN Secretary-General Antonio Guterres warned, "Storms, floods, fires, and droughts are decimating economies. Funds that should be building roads, educating children, and curing the sick are being swallowed up by the climate crisis."

The worldwide facts and figures are a shocker.

Hot On The Heels

The World Meteorological Organisation's (WMO) 'State of the Global Climate' report affirmed that 2023 broke records for being the hottest year for the concentration of GHGs in the atmosphere and for ocean heat content. Heatwaves reverberated through every facet of public health and the economy.

This was followed by the findings published in the journal Nature. The study revealed that the 2023 summer set a scorching record, marking the hottest season in the northern hemisphere in the past 2,000 years.

Researchers found that the temperature was at least 0.5°C warmer than the most extreme past climates, which means it was 2.07°C higher than in the pre-industrial period. The alarming trend was worsened by an El Nino weather pattern in the Pacific, intensifying the impact of human-caused global warming.

As climate change continues to be exacerbated by El Nino, temperature records are being broken again in 2024.

Severe drought and heatwaves swept across Europe. Southeast Asia sizzles under record heat waves.

The Philippines closed schools and warned of overloading on its power grid due to extreme heat. The temperatures in Thailand crossed 42°C in several provinces. After a long drought spell, Kenya, Ethiopia, and Somalia were hit by massive rainfall that caused devastating flooding last year.

The study found that the arid soil could not absorb the water and climate change doubled the rainfall intensity. Crop failures left millions of people facing food shortages.

Similarly, in many regions of the world, summers have become much hotter, increasing the risk of heatwaves, high-terrain wildfires, and unseasonal droughts. On the other hand, winters are wetter, snowballing into cloudbursts, flash floods, and rising sea levels.

The Cascading Effect

Investigators have pointed out the causes of the storm that hit the United Arab Emirates (UAE) and Oman in April 2024. The record rainfall flooded highways, inundated houses, grid-locked traffic, and trapped people in their homes. In the UAE and the Arabian Peninsula, where rainfall is rare, the temperatures soar above 50°C. Since it is one of Earth's hottest and driest desert regions, the UAE has led the effort to seed clouds and increase precipitation. But its meteorology agency proclaimed that no such operations occurred before the storm. The huge rainfall was due to a normal weather system exacerbated by climate change.

Likewise, despite normal to below normal temperatures over northwest India, extreme temperatures over the rest of the country, especially east and south peninsular India, have made the summer among the warmest since 1901. Contrarily, devastating floods in Pakistan affected 33 million people.

The destructive wildfires in the US and other countries are a wake-up call for all to be aware of and prepare for a new climate reality.

Friederike Otto, the senior lecturer in Climate Science at Grantham Institute of Climate Change and the Environment at London's Imperial College warned that hot spells would get more intense in the coming years. "Heatwaves are the deadliest type of extreme weather. They are often referred to as 'silent killers,'" she remarked.

Terming the 1.5°C temperature a 'political global

warming target', Otto reasoned that even if it were kept to 1.7°C, it would still be much safer than 2°C. However, she warned that our lethargy should not end up reaching 3°C.

Adding to the notion, in his book, 'Climate Change Isn't Everything', Professor Mike Hulme of the University of Cambridge, states, "It is quite easy to imagine worlds in which global temperature exceeds 2°C warming, which is 'better' for human well-being, political stability, and ecological integrity, for example, than other worlds in which – by all means and costs – the global temperature was stabilised at 1.5°C."

In 2007, Hulme received a personalised certificate from the Intergovernmental Panel on Climate Change (IPCC) recognising his contribution to the joint award of the Nobel Peace Prize to the IPCC.

The Finance Tussle

Even as the world looks up to policymakers to help keep the world safer from climate change, the primary agenda of the COP29 annual UN climate meeting at Baku, Azerbaijan, in November would be to negotiate a new goal for climate finance, as it seems, the term was not fully understood. The planners must focus on the source of the $3.6 trillion required to reach the net-zero goal.

Reading Between The Lines
The words 'climate finance' have been ill-defined since the flows largely compose debt, and barely 15 per cent is concessional. Nearly half of these are

from the private sector, and the contribution to climate adaptation that requires public finance has been unsatisfactory.

There needs to be clarity on the type of capital to be offered for adaptation and what will be available for mitigation. While private capital can be channelised to commercial mitigation projects, public finance must be dedicated to adaptation.

Still, it is felt that if the flow is in the form of debt, heavily indebted countries will stare at a dire fiscal future. Therefore, advanced countries must be urged to consider the macroeconomic constraints of the evolving nations. They must provide, if not a write-off, an option of a moratorium on the debt.

In the Ad-Hoc Work Program (AHWP) of the UNFCCC, held in Arizona in April 2024, the US stated that the New Collective Quantified Goal (NCQG) was 'voluntary' for those who 'chose to pay' under the 2015 Paris climate pact dealing with finance. It added that there should be a discussion on who would receive the money.

Understandably, developing countries decried that the goal must be according to the principles and provisions of the convention.

The Arab nations were forthright in stating that the $100 billion goal was in context with the Paris Agreement and there was unequivocal consensus of it being related to the developed countries. It stipulated that wealthy nations would provide financial resources to assist developing countries to mitigate and become adept at facing the impacts of climate change.

This was because the developed world was historically responsible for most emissions.

The emerging states maintained that the NCQG should be transparent on burden-sharing among advanced economies to seek accountability, as the latter's diversionary tactics were becoming obvious.

Solutions, Not Scars

It's a fact that much of the acceleration in the atmospheric concentration of GHGs is due to the pursuit of growth through globalisation and the world must acknowledge its adverse impacts. At this juncture, it would be wise to find responsible solutions, stop the blame game, and not scare the world about doomsday!

The remedy is to stop burning fossil fuels - oil, gas, and coal, and replace them with renewable energy sources. For, if the world keeps burning fossil fuels, the temperatures will continue to rise, inflation will increase, and the vulnerable will die.

On the other hand, adopting sustainable energy practices can help us step closer to building a renewable future. Major concerns such as oil shortage and fuel depletion can be solved, and the planet can be saved from future crises.

Experts propose adopting new formats, including

adjusting the working hours, altering school schedules, and running awareness campaigns to reduce distress.

Other important factors include better forestry management, adding green spaces, and sowing plants. Some strong allies in the fight against climate change are the trees, plants, and soil that store massive amounts of carbon at ground level or underground. When trees are cut, they release climate-warming carbon into the air and new plantations are not fast enough to replace them and require enormous time to serve the same function as old trees.

Therefore, responsible policies must end deforestation, as it destroys wildlife habitats and threatens the health and culture of Indigenous communities that live sustainably in green spaces.

In contrast, developing gardens, adopting regenerative agriculture practices, reducing pesticides, and rotational grazing help nurture the soil and produce healthy yields.

Planting mangroves in coastal areas offers significant climate change mitigation and adaptation co-benefits as they store carbon up to 400 per cent faster than land-based tropical rainforests. They protect cyclone-prone regions from rising sea levels, erosion, and storm surges, and are breeding grounds for marine biodiversity.

Transportation is a top source of GHGs, so eliminating pollution from the billions of vehicles driving across the planet is necessary. Governments are trying to phase out the sale of gas-powered internal combustion engine (ICE) vehicles and adopt electric automobiles.

Another important factor in mitigating GHG emissions is raising public awareness of cost and carbon-saving changes to buildings and making them energy-efficient.

The Plastic Paradox

Plastic is another dangerous polluter that poses a serious threat to life on Earth because of the toxic chemical additives used during production processes. Research reveals that it can enter the human bloodstream, permanently residing in the body, until death.

Every day, the equivalent of 2,000 garbage trucks full of plastic are dumped into the world's oceans, rivers, and lakes and 19-23 million tonnes of plastic waste leaks into aquatic ecosystems, polluting them.

Burning plastic and other types of waste releases dangerous substances such as heavy metals, persistent organic pollutants, and toxic chemicals into the air and stay in ash waste residues. These chemicals are linked to asthma, endocrine disruption, and cancer.

Therefore, until sustainable alternatives are found, we must reduce our dependence on plastic and keep it away from landfills and oceans.

In 2022, the UN General Assembly (UNGA) passed a resolution recognising the right to a clean, healthy, and sustainable environment as a human right. It called upon states, international organisations, businesses, and other stakeholders to strengthen the efforts.

Similarly, organisations like the National Environmental Scorecard in the US and Friends of the Earth in the UK urged political leaders to prioritise climate and nature action.

Still, the issue fails to gather attention compared to other concerns like inflation and employment.

Climate Trends polls revealed that only a small percentage of citizens viewed climate crisis as a major issue. So, being proactive on this front is a prerequisite.

As Jane Goodall, primatologist, anthropologist, and activist, remarked, "You cannot get through a single day without impacting the world. What you do makes difference, and you have to decide what kind of difference you want to make."

2

Fossil Fuels Burning Bright

If you think the environment is less important than the economy, try holding your breath while you count your money.

Guy McPherson,
Scientist, and Professor

Do you know that the Earth's climate was in natural balance two centuries ago? Trade between nations and continents was through sea and land, and mostly in perishable goods. The CO_2 in the air was stable at 280 parts per million (ppm).

Then, the Industrial Revolution spread its tentacles from Europe and North America to the rest of the world. Men began crossing boundaries by air, leaving a trail of CO2 emissions into the atmosphere. The industrial-embracing nations started burning coal and other fossil fuels to power their factories.

The two World Wars of the 19th century set machines rolling in countries to produce arms and ammunition for soldiers across the globe. These countries became richer and powerful, while others, especially in Asia and Africa, remained poor and underdeveloped, with little access to industry or technology.

A report by CNBC TV18.com on climate change provides a glimpse of the times gone by.

'When Scientists First Raised The Alarm

The concerns over climate change may have become bigger in recent years but the signs of brewing trouble were first given by the scientific community towards the end of the 1950s. Back then, it was previewed as a distant possibility in the 21st century for which the world should be prepared.

Live Science had reported quoting Spencer Weart, a historian and retired director of the Centre for History of Physics at the American Institute of Physics in Maryland. The concerns became serious after that, attaining larger scientific attention in the 1980s.

However, the history of people's concern over climate change may not date back to the last century only. The report claims that people's interest in understanding how human activities affect climate change dates back thousands of years.

*The report cites references from Weart's book, '***The Discovery of Global Warming***' to claim that the concerns over climate change existed in ancient Greece. From 1200 BC to AD 323, people in the region debated if draining swamps or cutting down forests would bring more or less rainfall. This was the first documented discussion around climate change and focused only on the local region.*

Weart claims that Swedish scientist Svante Arrhenius became the first person to imagine the impact of climate change globally on the existence of humanity in 1896. This was when he published his calculation in The London, Edinburgh, and Dublin Philosophical Magazine and Journal of Science, showing that more CO2 emissions would harm the planet. Arrhenius' work was then used as a base by the 19th-century scientist Joseph Fourier, who stated that the Earth would be a far cooler place without an atmosphere.

According to the Live Science report, other scientists of the era, John Tyndall and Eunice Newton Foote, used Arrhenius' work to demonstrate CO2 and water vapour-trapped heat, suggesting that an atmosphere could do the same. Though Arrhenius' precision proved largely accurate, it was not widely accepted then.

The scientific opinion around climate change concerns began in the 1950s due to two key experiments. The first of which was led by Roger Revella in 1957 showing that the ocean will not absorb all the CO2 released in humanity's industrial fuel emissions and that its levels in the atmosphere would eventually rise significantly.

Another study published by Charles Keeling in 1960 detected the rise in CO2 levels in the Earth's atmosphere. The first alarm around the impact of human activities was raised, followed by more studies tracing climate change to the existence of humanity. Towards the end of the 1980s, the scientific community finally acknowledged it as a problem that needed immediate attention.'

The Heat Is On

James Hansen, one of the most influential climate scientists of the era, warned that the world was heading towards a superheated climate not witnessed in a million years. Hansen spent several decades as director, NASA's Goddard Institute for Space Studies. He directs Columbia University's Climate Science, Awareness, and Solutions programme, and is at the forefront of decrying the lack of action to reduce emissions.

Hansen noted that the record heatwaves riling the US, Europe, and other parts indicated that the global warming experts had not communicated its ills properly. Another disappointing factor was that the elected leaders lacked the intelligence to respond to the issue. "It means we are damned fools," he remarked.

Eventually, at the 1988 Toronto Conference of the Changing Atmosphere, scientists and politicians converged to address climate change as a global threat to the Earth's atmosphere.

The National Oceanic and Atmospheric Administration (NOAA) Global Monitoring Laboratory data confirmed Hansen's alarm. It was revealed in mid-2022 that the global average CO2 concentration was 421 ppm, the highest level since the mid-Pliocene, two to four million years ago.

The concentration rose by 50 per cent since the Industrial Revolution when it was around 280 ppm. The main causes were the burning of fossil fuels, changes in land use, and deforestation.

Burning fossil fuels meant emitting CO_2 and other GHGs, which trapped the heat in the atmosphere that would otherwise leave from Earth to outer space. Known as the greenhouse effect, it is essential for life on the Earth, but too much of it causes the planet to heat up too rapidly, enhancing the reaction and resulting in climate change.

It became evident that human activities that cause severe social problems and exacerbate environmental issues comprise industry, agriculture, deforestation, transportation, and electricity generation. Though these catalyse global trade, commerce, and innovation and elevate the quality of life for billions, the pursuits also release GHGs and lower the ability of the atmosphere to absorb them.

They are the largest contributors to global warming.

A large chunk of the GHGs blanketing the Earth and trapping the sun's heat are generated through energy production by burning fossil fuels to generate electricity and heat. Fossil fuels account for 90 per cent of all CO_2 emissions and over 80 per cent of global energy production. These have powered the industrial growth and development of civilisations for centuries.

The notable early discoveries include China's use of natural gas in 500 BC and Persia's utilisation of oil in

600 BC. In North America, indigenous people used oil before the modern industry took off in 1859.

A report by CNN on the US coastline:

'Dozens of cities along the US coastline are sinking at alarming rates, leaving them far more exposed to devastating flooding from sea level rise than previously thought. As oceans rise and coasts sink, up to 343,000 acres of land will be exposed to destructive flooding by 2050 from hazards such as hurricanes, coastal storms, and shoreline erosion, according to a study published in the journal Nature. In a worst-case scenario, roughly 1 in 50 people in the 32 cities analysed could be exposed to flood threats.'

Furthermore, oil spills, gas leaks, and fracking inflicted severe damage on the ecosystem, resulting in public health. The extreme challenges became a source of geopolitical strife, leading to wars and conflicts between neighbouring countries over their control and distribution.

The finite and non-renewable nature of fossil fuels has left humanity facing a daunting deadline. Predicting the exhaustion of these resources is a complex endeavour, influenced by reserves, consumption rates, technological advancements, and price fluctuations.

According to Octopus Energy, 2060 may mark the twilight of fossil fuels based on current consumption and reserve estimates.

Oil And Gas, An Exhausting Issue

In 2020, global oil production reached 91.7 million barrels per day, with players like the US, Saudi Arabia, and Russia contributing significantly. Simultaneously, natural gas production stood at 3.94 trillion cubic metres.

Experts have warned that to avoid the worst impact of climate change, emissions must be reduced by half by 2030 and net zero by 2050. According to the UN, "To achieve this, we need to end our reliance on fossil fuels and invest in alternative sources of energy that are clean, accessible, affordable, sustainable, and reliable."

However, the sustainability of eco-friendly alternatives rests upon complex factors. There's an intricate interplay between these resources and the innovations needed to ensure a viable energy landscape.

Meanwhile, the UN climate chief Simon Stiell recently warned that the world has only two years left to avert climate catastrophe. During his April 2024 address at the Chatham House in London, he urged the countries to strengthen their climate plans, known as nationally determined contributions (NDCs) under the Paris Agreement. Stiell cautioned that current NDC goals would barely reduce emissions by 2030.

Nevertheless, his statement has a silver lining. "We still have a chance to make GHG emissions tumble with a new generation of national climate plans... But we need these stronger plans – now. And while every country must submit a new plan, the reality is that G20 emissions are around 80 per cent of the global emissions."

The Knowledge Curve

Learning from the mistakes, as the world works towards increasing renewable electricity to reduce carbon emissions, the International Energy Agency (IEA), a Paris-based autonomous intergovernmental organisation, estimates that the Earth will add 305 GWs of renewable electricity annually from 2021 until 2026. This could save about 683 million tonnes of carbon emissions yearly if renewable energy has a 30 per cent capacity common for wind and solar power.

However, this could be negated due to carbon emissions from wildfires, which have gone from bad to worse. Wildfires release millions of tonnes of carbon in a single season.

GLOBAL WILDFIRES

CARBON ELECTRICITY

On The Wild Side
A conservative estimate reveals that wildfires could emit 12 gigatonnes of carbon in the next

three decades. In August 2021, global wildfires emitted 1.3 gigatonnes of carbon dioxide, mostly in North America and Siberia. This was almost double the carbon savings from renewable electricity in a year.

The Challenge of Our Times

Beginning in March 2023, with increased intensity starting in June of that year, Canada faced a record-setting series of forest fires, affecting all 13 provinces and territories, including Alberta, British Columbia, Ontario, and Quebec. In late June, the smoke crossed the Atlantic, reaching Europe. It surpassed the 1985, 1995, and 2014 wildfire seasons. By early October, about 5 per cent of the forest area was burned, and of the 772 active wildfires, 346 were deemed 'out of control'. Smoke emitting from the fires caused air quality alerts and evacuations in Canada and the US.

Nepal also sees a spate of wildfires annually, and the numbers and intensity have worsened. Over 4,500 wildfires have been reported this year across the country. As a result, temperatures have risen above 40°C in the Buddhist pilgrimage city of Lumbini and other parts.

What Causes Forest Fires?
Occurring due to a rise in temperatures, the presence of a huge quantity of dry leaves acts as a catalyst.

The Economic Impact

Forest resources, including carbon locked in the biomass, are lost in these fires. According to the World Economic Forum, forest fires cost $50 billion annually, greatly impacting real estate and tourism.

High-Voltage Spectacle

The 2022 report of the Intergovernmental Panel on Climate Change (IPCC) states that the severe impact of climate change is expected if global temperatures rise 1.5°C by 2030:

- The heatwaves will affect at least 14 per cent of the population globally – at least once every 5 years – resulting in respiratory and cardiovascular diseases, diabetes mellitus, and renal disease. If the Earth warms just 1°C from the current target of 1.5°C, wildfires will burn almost 35 per cent more land globally than the current damage.

- Rising seas will submerge tens of millions living in coastal areas, and small islands will vanish. For example, the atolls in Maldives, Kiribati, Tuvalu, and the Marshall Islands face the highest risk. About 1 billion people will face coastal flooding from rising seas. This would mean that many of them have to leave their homes. Some islands will be uninhabited if the global temperature rises – a few more tenths of a degree than the current goal temperature.

- Droughts and floods will result in food scarcity and uneven yield production. In India, floods and droughts damage crops, resulting in food inflation and rising prices. In Sudan, flooding disrupted the planting season and the conflict in the region prevented families from accessing farmland and driving hunger and malnourishment. In Somalia, rainfall has been below adequate levels for harvesting and keeping livestock and people are experiencing the worst form of hunger. The floods affected millions in Pakistan and its malnutrition rates are dire.

- Arctic ice thaws are expected and 40 per cent of its permafrost will be affected severely by the end of the century. A NASA map report shows the minimum size of the Arctic Sea ice measured since the satellite observation started in 1979, has depleted, and the 2012 sea ice extent is the lowest in the satellite record.

- Various plant species, animals, insects, and other living organisms will be at risk of extinction. The polar bear population declined in Alaska and Canada, while 65 per cent of the insect population could be extinct over the next century.

Likewise, the impact and the damage could be greater if it reached the 2°C threshold.

The IPCC report stressed that among the 13 developing countries with large energy consumption, 11 were exposed to high energy, security, and systematic risks. Therefore, they were forced to seek damage support from developed nations.

For instance, though Africa has contributed less to GHGs, it has suffered widespread loss and damage to its biodiversity, food production, and lives. By 2030, less than 100 million people in the continent will be exposed to rising sea levels, compared to half of it 20 years ago.

By then, the climate impact will cost developing nations between $290 billion and $580 billion annually. It will double by 2050. Though the US, the European Union, and Japan are historically the world's biggest polluters, China and India have joined the list due to rapid industrialisation and urbanisation.

Lack of Initiatives

On the eve of COP27, the Emissions Gap Report highlighted that countries with climate commitments were slack with their implementation, putting the world on a trajectory for a disastrous 2.8°C rise in the temperature by the end of the century. Even if the current pledges were met, the world could still face a 2.4°C to 2.6°C rise in temperature by the end of 2100. Thus, the future looks bleak, as at least 3.3 billion lives would be affected.

Reports indicated that people were 15 times more likely to die from extreme weather conditions than natural causes. Between 2030 and 2050, climate change will cause an additional 250,000 deaths annually.

High temperatures increase the ozone level and other pollutants, which lead to 1.2 million deaths annually due to cardiovascular and respiratory diseases.

Thus, it is crucial to demonstrate solidarity and offer a collective response to those hit by climate change.

Mitigation and Adaptation

In climate change literature, 'mitigation and adaptation' means combating climate disasters. Mitigation involves reducing the heat flow and trapping GHGs into the atmosphere to cut the impact of potent gases from interfering with the Earth's atmosphere. Adaptation means adjusting to the expected disasters to reduce the risks of climate change. However, adaptation and its impacts are expensive and can go up to $140 billion to $300 billion annually.

As part of the mitigation and adaptation techniques, it is to be noted:

- *GHGs have the potential to warm the atmosphere. Each gas has a different amount of warming and duration of being active. The global warming potential is a term used to indicate the amount of warming in a specific duration (normally 100 years). If these gases were replaced or removed from daily usage, it would achieve a remarkable decline.*
- *The environment and construction industry represents about 40 per cent of the GHG emissions. Building, constructing, and managing it should be considered a major source of potent gas enablers. Therefore, environmental social governance should be mandatory. The rational use of materials and services such as water, heating, air-*

conditioning, and fuel consumption within the
industry will help.

*   As part of the finance scheme agreed upon in
    COP27 in Egypt, a certain budget should be
    allocated to research institutions globally to
    combat the issue.

The Science of The Earth's Layers and Heat Exchange.

Some crucial scientific points to be understood include
the planet's atmosphere which comprises layers of
different characteristics.

The troposphere, the lowest layer, extends about 10
km above sea level. The next layer, the stratosphere,
extends about 50 km above ground level. The third,
mesosphere, extends to a height of 85 km. Beyond this,
the thermosphere is a layer of rare air.

These atmospheric layers have different heat exchange
abilities. The troposphere and stratosphere transfer air
and atmospheric constituents across the tropopause.
The tropopause and the lower stratosphere layers are
crucial in shaping the Earth's atmosphere. On leaving
the ground, there is a reduction in air pressure, density,
and temperature. The varying temperature is the
determining phenomenon of the atmospheric
structure. Most clouds and weather changes appear in
the troposphere in the first seven miles.

Over the last three decades, scientists and researchers
have investigated the ability of people to warm up the
environment. With the help of the UN organisations,

though the most potent gases and their abilities were recognised, limited achievements materialised in reducing global warming.

The temperature continued to rise, impacting lives. So, whether the boundaries of the Earth's atmosphere are fully understood requires answers, and if heat exchange between the lowest two atmospheric layers has ramifications. Or are we missing the ability to change the temperature of the atmosphere?

Human presence exists in the atmosphere's lowest layer, where all weather conditions prevail. About 99 per cent of water vapour exists here because most clouds are generated within the troposphere. On moving higher, air pressure drops, and temperatures become colder.

This is unlike the stratosphere where the ozone layer exists and gets warmer at high altitudes. The rising temperature indicates turbulence. The phenomenon of having two layers with different temperatures and variations leads to possibilities of heat exchange. The troposphere is mainly heated by ground contact and not directly by solar radiation.

Heat transfer by convection results in mixing and heating the troposphere. It is responsible for air scaling up and becoming less dense amid buoyant and turbulent bubbles. Due to this process, the temperature decreases, expanding isolated bubbles of air heated by the Earth and moving up.

The bubbles continue to rise until the temperature matches the air in the vicinity. The balanced state of the bubbles and their surrounding is the boundary between the troposphere and the stratosphere. It acts as a barrier – hindering, mixing, and transferring heat by convection.

The tropopause has limitations in sealing and preventing heat exchange between the layers. The existing knowledge identifies processes that can transfer heat and air across the tropopause.

One way of the transfer happens when thunderstorms and tropical cyclones penetrate the tropopause and inject water vapour and heat into the stratosphere. The other takes place due to gravity, where atmospheric waves are generated by mountains, wind shear, and convection.

This disturbance propagates upward and transfers the heat in the air. The planetary waves in the atmosphere, generated by the rotation of the earth and the distribution of sea and land, can also break in the stratosphere and cause modification in the temperature. The internal exchange between the stratosphere and the troposphere adds to it and transports the air and atmospheric constituents across the tropopause barrier.

The heat exchange between the troposphere and the stratosphere can impact climate change significantly. It affects the system, altering the radiative, balance, and feedback between the two layers, which means the Earth's climate is impacted by heat exchange between the atmospheric layers.

Nevertheless, for now, we are looking for progression, not perfection. It is, therefore, important to discuss other relevant and worrisome factors.

3

Construction Industry's Accountability

Just as engineering marvels under the aegis of the construction industry take years to complete, so does time spent securing our goals in life.

Anonymous

The COP28 held in Dubai, in 2024, was momentous in concluding the first global evaluation of the world's laborious effort to address the all-encompassing goals of the international treaty to confine the global average temperature to below 2°C.

At the conference, it came to the fore that progress was at a snail's pace across all climate areas, including reducing GHG emissions, strengthening resilience to a changing climate, and procuring financial and technological support to vulnerable nations. The participating countries, however, unequivocally pledged to accelerate actions by 2030.

The Summit also noted that while adopting ambitious targets to meet an individual nation's demand, it was time to point out that one of the main drivers of environmental degradation in the world was the robust construction industry.

Though thrust is given to construction companies to thrive and survive through development, improving the planet's health is as pertinent as nursing a patient with a high fever. Thus, the UAE raised a red flag and warned the IPCC that it was way off-track in meeting its objective.

Largest Polluter

The building and construction sector is the largest emitter of GHG, accounting for 37 per cent of total global emissions due to cement, steel, and aluminium production. As per the IEA findings, it accounts for 23 per cent of air pollution, 40 per cent of drinking water contamination, and 50 per cent of landfill waste. It is also responsible for 30 per cent of energy consumption and 27 per cent of the total energy sector emissions.

Amid growing public awareness and governments' initiatives to reverse carbon emissions, the clock is ticking for the construction sector to put its act together on redefining its role in taming the ecological imbalance.

'Dzud' Facts In Mongolia

The ongoing white and iron dzud in Mongolia has reached a critical level, with over 90 per cent of the country facing high levels of risk from the unique weather phenomenon, the UN reported.

About 190,000 herder households are struggling with inadequate feed, skyrocketing prices, and heightened vulnerabilities, as per the office of the UN Resident Coordinator in Mongolia.

Herding and livestock have traditionally been integral to the country's economy, culture, and way of life. Estimates indicate that over 64 million livestock will be affected this winter due to the increasing severity of weather conditions, exacerbating the crisis.

It underscores the urgent need for humanitarian assistance and sustainable solutions to support Mongolia's rural communities and traditional livelihoods.

According to the UN Economic and Social Commission for Asia and the Pacific (ESCAP), the frequency and intensity of dzuds have increased since 2015 due to the worsening impacts of climate change and poor environmental governance.

What are dzuds?
Dzuds, a peculiar slow-onset disaster unique to Mongolia, are extreme winters characterised by freezing temperatures, heavy snow, and frozen ground where animals cannot reach the pastures.
These conditions are commonly preceded by a dry summer with equally scant grazing; leaving livestock unable to build up the stores of fats needed for winter.

The Point Mark

To keep the 2030 world's climate target within the 1.5°C limit, the IEA advocated 20 per cent of all existing buildings to be renovated by stressing the construction industry's decarbonisation effort.

The autonomous intergovernmental organisation sought an annual 'deep renovation rate' of over 2 per cent until 2030. However, in cases where constructing a new building is essential, urban planners must, at the outset, ensure the implementation of eco-friendly green designs.

The architecture and the impact of structure on inhabitants must be planned for long-term environmental benefits by using less land and focusing on connectivity through public transport, walkways, and cycle tracks to discourage personal vehicles from plying on roads.

Increased access to green and blue spaces such as parks and water bodies will improve air quality and cocoon natural resources and boost the health and well-being of the populace and ecosystem.

As the inclusion of Environmental Social Governance (ESG) standards within the industry becomes increasingly important, pressure is mounting on the construction sector to clean up its act.

Setting the ESG standards must be mandatory globally to evaluate the sustainability of new developments and renovated structures.

There should be an urgency to galvanise ways, so that, instead of new constructions that use huge resources, a mechanism is adopted to renovate existing edifices with robust bases. After all, refurbishment improves insulation and reduces energy usage.

As American musician David Allen Coe once said on concrete, 'It is not the beauty of the building you should look at; it's the construction of the foundation that will stand the test of time!'

Leading By Example - Energy-Efficient Structures

Contextually, an important aspect is the use of passive design combined with renewable energy, which significantly lowers the carbon footprint of an inhabited building.

It begins by opting for suitable building locations, designing the layout of the rooms and windows, and taking measures for proper insulation, ventilation, thermal mass, and water harvesting systems, which play crucial roles in making an energy-efficient building.

Compared to a conventional structure, a certified passive house uses about 80 per cent less energy to heat and cool. By adding solar panels or wind turbines for power generation and water heating, the energy demand and its environmental impact can be lowered considerably.

The new generation photovoltaic solar tiles guarantee even greater flexibility and enhanced returns on financial investment. Additionally, using geothermal and air-source heat pumps increases the efficiency enormously compared to traditional gas boilers.

Lifestyle Modification

When India's Maharashtra experienced intense heat this summer and people switched on their air conditioners, architect Rohan Nahar, who runs a startup Studio Roots and Basics, needed nothing more than a fan in his Pune office! It is built with bare hands, using construction waste, lime, and (available on the site) 'mud', which he refers to as 'brown gold'.

Since 2013, his construction pillars have been environmental, economic, and social sustainability. Nahar has collaborated with government departments to incorporate basic architectural principles - natural light and ventilation. The man makes sense when he says, "If you sit under a tree and are comfortable, but if you sit inside your house and are not comfortable, there's something wrong."

He has modelled several structures using mud, an old tradition in India and many parts of the world. Nahar is much sought after by people and organisations from the US, Spain, and Portugal who desire to create homes with natural materials to reduce their carbon footprint.

Say No To Concrete

Concrete, a building material comprising a mix of cement, water, sand, and rock, is extensively used worldwide. It releases more than 4 billion tons of CO_2 annually. Cement – the binding agent made from limestone and clay – is described by sustainable building experts as the 'most destructive material on earth'.

The kiln process of cement manufacturing involves the emission of large amounts of CO_2, which leads to global warming. Its production vents up to 8 per cent of global CO_2 emissions.

The cement industry contributes about 5 per cent of global anthropogenic CO_2. This is mainly because the production process is highly energy intensive and involves calcination of raw material and fuel combustion. It is used to construct roads, buildings, bridges, and dams.

China is the largest producer and consumer of cement. The Three Gorges Dam on China's Yangtze River is the largest concrete dam, which took 17 years to build. It is 185 metres high and 2309 metres long.

History Of Cement

The use of cement dates back thousands of years to ancient civilisations. The Romans invented non-hydraulic cement by heating limestone, producing calcium oxide and CO_2. However, since it was soluble in water, it was unsuitable for high-moisture environments and often resulted in the collapse of the structures.

Comparatively, hydraulic cement was durable and

stable in water; thus, it was used to construct various historical sites, which still exist.

However, since volcanic ash was rare and the setting time took long, there was a requirement for better hydraulic cement.

About 1800 years later, England and France replaced volcanic ash with clay, shale, and slate, though the setting time remained high. Finally, in 1824, Joseph Aspdin of England invented and patented Portland cement, which developed into the 'modern Portland cement' by his son William Aspdin.

In the various stages of cement manufacturing – from extracting raw material to crushing and kiln phase to final grinding – leading cement manufacturers are now involved in extensive research to reduce or eliminate these emissions. This is possible by incorporating recycled building materials, including fly ash and slag, which have a much lower detrimental environmental impact.

Another alternative is hempcrete, created from hemp plants mixed with a lime-based binder. This forms a lightweight, breathable construction material with excellent insulation properties. The wholesome option is rammed earth, generated by compressing soil into a formwork. It is durable, low-maintenance, and has excellent thermal mass properties.

Modern breakthroughs in construction materials include straw bale and cross-laminated timber. Though produced with low environmental impact, these match the existing construction ingredients in terms of practicality and strength.

Though living in a futuristic world is our aspiration, it should not come at the expense of 'Mother Nature'. We have reached a point, were; by pulling the blinds of our comfort zone, we will not be able to shy away from the ills of global warming.

4

The Economic Sense of Green Energy Resources

Our carbon emissions must eventually go to zero. Otherwise, we are never going to have a stable climate and that's what our goal is for human civilisation to thrive, a stable climate.
Katherine Hayhoe,
Canadian Atmospheric Scientist

None will deny that even the most complicated issues can be resolved if the world is treated as a family. The solutions are around us; the need is to open our eyes. This is akin to renewable or green energy sources that are plentiful and all around us. Unfolding and channelling them can turn the tables.

Solar energy from the sun, geothermal energy from heat inside the Earth, biomass from plants, hydropower from flowing water, wind power, ocean tides, and hydrogen hold the key to displacing finite energy sources like fossil fuels and nuclear power, which harm the environment and the health of living beings.

Currently, about 29 per cent of energy comes from renewable sources. Lauded for their low environmental impact and global accessibility, their sustainability hinges on multifaceted variables, including consumption rates, availability, social consequences, and technological breakthroughs. The path to securing these for future generations is immediate, requiring prudent management.

To measure the progress of renewable energy, we must know its generation and output. While generation means the amount of electricity produced from renewable sources, output denotes its amount in the total electricity mix – including fossil fuels and nuclear power.

The IEA estimates that renewable electricity generation will grow by 60 per cent by 2026, reaching over 10,000 TWh annually. The output will increase only up to 34 per cent since fossil fuels and nuclear power will still dominate the electricity mix.

That's because electricity is a part of the total energy consumption, including transport, heating and cooling, and industrial use. The ratio of its generation to production depends on how it is measured. The methods are direct consumption, where fossil fuels like coal, oil, and gas are used, and substitute consumption, where nuclear, solar, wind power and ethanol are used.

The direct system counts the primary energy sources without correcting the fossil fuel inefficiencies and biomass conversion. The substitute method corrects these inefficiencies by converting nuclear and modern renewable sources to their equivalent primary energy from fossil fuels.

Under this method, in 2019, the global ratio of electricity generated was 18.4 per cent, whereas, as per the substitute process, it was 22.8 per cent. Electricity accounts for less than a quarter of the world's energy production.

A report in Statista says global energy consumption has increased and by 2050 renewable energy consumption will rise to 247 petajoules. Comparatively, in 2000, the total renewable energy consumption was 42 exajoules.

What is an exajoule?
It is the International System Unit of electrical, mechanical, and thermal energy. A unit is equal to the work done when a current of one ampere is passed through a resistance of one ohm for one second. A unit of energy equals the work when the force of one Newton acts through a meter.

Though renewable energies are inexhaustible, they pose a great challenge for governments to sustain them.

However, achievements by different countries are boosting the morale of other nations.

The Pros And Cons

Solar and wind, though considered the cheapest forms of electricity, produce power only when the sun is shining, or the wind is blowing. The rest of the time, backup systems are needed. This makes electricity expensive and that's the reason global electricity remains almost two-thirds reliant on fossil fuels. By this reasoning, experts claim it may take a century to eliminate fossil fuels.

Bjorn Lomborg, the president of the Copenhagen Consensus and visiting fellow at Stanford University Hoover Institution stated, "Modern societies need power 24x7, so unreliable and intermittent solar and wind sources entail large and often hidden costs. This is a smaller problem for wealthy countries that have already built fossil-fuelled power plants and use them as a backup. It will, however, make electricity more expensive, as intermittent renewables make everything else intermittent too."

He rightly pointed out that advanced economies refuse to fund profoundly needed fossil fuel energy in the developing world. These countries insist that people cope with unreliable green energy supplies, which, unfortunately, cannot even power water pumps or agricultural machinery to lift the population out of poverty. Lomborg refuted the claims that emerging industrial powers like India, China, and Indonesia derived more power from solar and wind. He maintains that these countries get additional power from coal.

Importantly, the cost of solar and wind energy is lower when the sun shines and the same goes for the wind. However, it becomes exorbitant due to unreliability and the requirement for electricity storage in the absence of the sun and the wind.

The president of the Copenhagen Consensus stated that the low-cost claim of solar and wind energy was floated, as it was easy to do so. He suggested investing in low-carbon energy research and development (R&D), which could bring a technological breakthrough in related matters and other technologies like modular nuclear power. Meanwhile, some nations are setting a precedent.

Small farmers adopt bio-digesters in Africa
Statistics from Energy Capital and Power, an Africa-centric global investment platform for the energy sector stated that Kenya, a flag bearer in biogas policies, is heading towards establishing 17,000 household biogas digesters, and 8,000 biogas plants. Furthermore, Tanzania, a leader in Africa's biofuel potential countries has 12,000 digesters. Uganda, which launched a four-year Africa Biodigester Component Project in 2022, has 7,000 domestic biogas plants, according to the Uganda National Biogas Alliance.

India empowers village with solar power
Modhera, located in Gujarat, has become the country's first solar-powered village, drawing praise from all over, including the UN Secretary-General António Guterres, who called it a symbol of

"reconciliation between humankind and the planet."

Set up with an investment of $9.7 million, the solar initiative has changed the lives of the residents, enabling them to save substantially on their power bills.

China races ahead in renewable energy

The IEA in its Renewables 2023 report said that during 2023-2028, China might account for 56 per cent of renewable energy capacity additions. The country will increase its renewable capacity by 2,060 gigawatts (GW), while the rest of the world will add 1,574 GW. The European Union and the US are the next biggest renewable energy builders at 429 GW and 337 GW, respectively.

Thus, sustainable options have become essential to energy discussions. The quest for sustainability is no longer optional; it is fundamental to securing a prosperous future.

Scarcity vs Sustainability

Minerals like copper, nickel, manganese, cobalt, lithium, graphite, and rare earth elements (REEs) are the driving force of clean energy technologies, powering solar photovoltaic modules, wind turbines, batteries, electrolysers, and fuel cells. However, their distribution is uncertain due to geological scarcity, environmental and social repercussions, and market dynamics.

The IEA revealed that the supply and investment plans for critical minerals fell far short of the demand to accelerate the deployment of solar panels, wind turbines, and electric vehicles.

The Sustainable Development Scenario (SDS), aiming for net-zero emissions by 2070, anticipates a staggering 4 to 6-fold increase in mineral demand by 2040. By then, the minerals required for clean energy technologies will double if climate change policies are moderate. It can be given a four-fold stimulus if the policies are ambitious.

However, it does not denote the presence of enough minerals to produce renewable energy. Their supply usually depends on resolving geopolitical issues and ensuring international collaboration.

A significant source of the minerals used in the manufacture of renewables is unearthed in China. However, as political equations stand between Beijing and the West, the latter will presumably be reluctant to provide the world's second-biggest economy with a clear path to boost its growth.

In several other countries, essential mineral mining often requires adherence to complex legal processes. For instance, obtaining minerals from Africa and parts of Asia is not as simple as 20 years ago. According to the US Department of Energy, global energy demand will double by 2040 due to rapid economic growth in China, India, Indonesia, Brazil, and the Middle East.

Therefore, generating electricity or heat from renewable sources instead of fossil fuels reduces the countries' dependence on imports and vulnerability to geopolitical shocks and crises.

According to the International Renewable Energy Agency (IRENA), green energy sources are cheaper than fossil fuels in most parts of the world as their cost will diminish further due to technological innovations and economies of scale. These can boost economic growth, create new jobs, reduce poverty, and provide social justice for everyone.

The recent price surge has ignited investor interest and skepticism in the energy transition. This 'greenflation' highlights the advantages of renewable materials, minerals, and technologies to global warming mitigation. While strategies exist to mitigate supply risks, it is essential to comprehend the gap between mineral scarcity and sustainability.

Can't Dig Deep

In its recent report, the American multinational strategy and management consulting firm, McKinsey & Co. alerted on the impending shortages. It gauged a potential 20 per cent shortfall in nickel and a discouraging 70 per cent deficit in dysprosium, a rare earth element crucial for electric motors. The study listed copper, nickel, and rare earth elements as the most favoured minerals, followed by lithium and graphite. It stressed that the shortage could impede global decarbonisation efforts, leading to price fluctuations and volatility.

Similarly, the United States Geological Survey (USGS) provided estimates of the world mineral reserves essential for renewable energy technology. Irrespective of the extent, minerals critical to renewable energy are finite and inherently unsustainable.

However, a major barrier is the lack of adequate policies, regulations, incentives, and financial support for the clean transition. Another is the inconstancy of renewables such as solar and wind, which require large battery storage facilities, allowing them to integrate successfully with existing systems and grids. The social and environmental impacts, such as conflicts over land use, biodiversity loss, and water consumption, are other crucial challenges.

Green energy production can supplement fossil fuels only if there's a historical and systematic approach. This calls for educating societies on the value of energy saving and providing incentives to individuals for their health initiatives.

It also requires adequate funding to global research institutions and universities to support reliable and science-based solutions.

Departure From Clichés

It's time to approach the issue holistically, extending it beyond energy source considerations to encompass the technical facets of renewable energy production. There's a need to understand that if the minerals driving these technologies are not guaranteed, the sustainability of natural energy sources like sun, wind, and water cannot be sustained.

For better outcomes, the IEA prescribed a series of actions, including bolstering mineral data and information, fortifying supply chain resilience and transparency, promoting responsible mining practices, scaling up recycling and material efficiency, and fostering innovation and diversification.

The answer lies in innovation, which has emerged as a viable strategy to reduce mineral dependence. Analysts have indicated that these materials may not endure in a net-zero scenario without substantial mitigation measures. Therefore, transitioning to technologies with reduced mineral requirements could mitigate sustainability concerns. The innovation-driven shift towards less reliant mechanisms will provide a viable solution.

While the current reserves may meet renewable energy demands shortly, the ongoing supply challenges and risks require vigilant monitoring and analysis.

In such a scenario, exploring alternative sources alongside diversification and innovation is pivotal to safeguarding supply security and sustainability in the quest for a greener future.

As American historian and philosopher Howard Zinn said, "We don't have to engage in grand, heroic actions to participate in change. Small acts, when multiplied by millions of people, can transform the world."

5

Banking On Green Hydrogen

Being challenged in life is inevitable, being defeated is optional.

Roger Crawford,
Tennis player and motivational speaker

It is said that whatever the world can do, the UAE can do better!

The Emirates produces green hydrogen at the Mohammed bin Rashid Al Maktoum Solar Park. Green hydrogen is the most environmentally friendly way to produce hydrogen, which is a clean and versatile energy carrier with various purposes, including electricity generation, transportation, and industrial processes. It does not emit carbon or other pollutants.

However, as fuel, it faces several challenges and barriers, such as high production costs, lack of infrastructure, policy uncertainty, and safety issues. That's because most of the hydrogen produced comes from fossil fuels, such as natural gas and coal, which emit GHGs and contribute to climate change.

Even when hydrogen is used in fuel cells, it emits nitrogen oxide (NOx) that includes two gases – nitric oxide (NO), a colourless, odourless gas, and nitrogen dioxide (NO_2), a reddish-brown gas with a pungent odour. NO reacts with oxygens or ozone in the air to form NO_2, which harms human health and the environment.

One pound of nitrous oxide (N_2O), a type of NOx, can cause 265 times more global warming than one pound of CO_2. Hydrogen also generates pollution when it is burned, as it produces more NOx emissions than natural gas. Therefore, hydrogen is not a clean fuel if manufactured without the help of renewable energy.

Therefore, with the focus on green hydrogen, many countries are investing in it as a potential solution to reduce their dependence on fossil fuels and achieve their net-zero emission targets.

The Gulf countries are among them, as they have abundant solar resources that can produce green hydrogen through water electrolysis.

As per the World Economic Forum report, a UN initiative – 'Green Hydrogen Catapult' – will boost hydrogen energy for green electrolysers from 25 GWs in 2020 to 45 GWs by 2027.

According to Dr. Emanuele Taibi, head of the Power Sector Transformation Strategies at IRENA, green hydrogen is a chemical element produced by splitting water into hydrogen and oxygen using renewable electricity.

This is a versatile path compared to both grey and blue hydrogen. Grey hydrogen is produced from methane (CH_4), split with steam into CO_2 – the main culprit for climate change.

While blue hydrogen follows the same process as grey, with additional technologies necessary to capture the CO_2, it is produced with hydrogen split from CH_4 or coal and stored for a long term.

A BBC report stated that Saudi Arabia plans to use hydrogen as a new fuel source. Since it will be extracted from water using renewable energy, the hydrogen will be green. The process involves using electricity from solar or wind power to separate hydrogen and oxygen molecules in water. While exported to other countries, green hydrogen will power Neom, a futuristic city costing $500 billion.

Oman also aims to build the world's largest green hydrogen plant in Duqm to transform it into a leader in renewable energy technology. Its construction will begin in 2028 in Al Wusta and is aimed at full capacity by 2038. The plant will be powered by 25 GWs of wind and solar energy.

The consortium behind the $30 billion project includes oil and gas company OQ, Hong Kong-based InterContinental Energy, and Kuwait-based energy investor Enertech.

Annually, the facility will produce 1.8m tonnes of green hydrogen and up to 10m tonnes of green ammonia. The maximum production will be exported to Europe and Asia.

Oman relies heavily on fossil fuels, generating up to 85 per cent of its GDP from oil and gas. Its fossil fuel reserves though are dwindling and becoming increasingly costly to extract. The country hopes to diversify the economy away from these and increase investment in renewables by 2040.

The IEA called for ending fossil fuel investments if governments were serious about climate commitments.

Since one of the stumbling blocks for green hydrogen has been cost, partly because of the massive energy required, according to the Wood Mackenzie research, as renewables and electrolysers become cheaper and fossil fuel prices rise, the cost will fall to 64 per cent by 2030.

"Most green hydrogen products will not be competitive for at least another decade," said Falko

Ueckerdt, a senior scientist at the Potsdam Institute for Climate Impact Research.

Meanwhile, countries like Australia and Brazil are also geared towards the green hydrogen mission.

Scientists claim that Gulf and other oil-producing countries should be aware of the limitations and risks of relying too much on hydrogen as an alternative to fossil fuels that are finite resources and will eventually run out.

According to estimates, oil, gas, and coal will be exhausted by 2072, 2075, and 2136, respectively, depending on consumption patterns and the discovery of new reserves. Therefore, the world needs to prepare for a post-fossil fuel era, as hydrogen is not a primary energy source, but an energy carrier, produced from other sources. It is not always efficient or optimal to convert renewable energy to hydrogen and then use it as a fuel, if possible, direct usage is better, such as in electric vehicles or grid-connected applications. Hydrogen must be expended as a complement to electrification, not as a substitute. It needs to be prioritised for applications where direct electrification is not possible or practical, such as heavy industry, long-distance transport, and seasonal energy storage.

Importantly, hydrogen is not yet a mature technology; it demands more R&D to reduce costs and increase efficiency. The technologies and solutions must overcome challenges, including the durability and performance of fuel cells and electrolysers, developing new materials and catalysts, enhancing storage and distribution systems, and ensuring safety standards and regulations.

Hydrogen must compete with other low-carbon technologies and solutions, such as batteries, biofuels, synthetic fuels, nuclear power, carbon capture and storage, and utilisation.

The Gulf region must fathom that hydrogen is not an antidote that can solve all their energy problems. It is a promising option that can play a key role in decarbonising some sectors of the economy that are hard to electrify.

Along with diversifying their energy portfolio and exploring different options for sustainably producing hydrogen, they should also continue relying on solar power and other renewable sources that can provide clean and abundant energy for their needs and for exporting green hydrogen to other regions.

Is Blue the New Green?

As green hydrogen is being marked as a reliable, versatile, and clean source that offers a solution to the climate crisis, it is hugely energy-intensive and presently too expensive to be commercially feasible. Therefore, most hydrogen produced is blue and two to three times cheaper. The question thus is, should blue hydrogen have a place in a decarbonised energy future?

The distinction between the two is substantial. While blue hydrogen does not produce any emissions when burned, its production is far from clean. Blue hydrogen is extracted from natural gas in a process that requires a lot of energy and emits vast amounts of CO_2.

While the purpose is to capture and store the gas underground, there are concerns about the stored supply of gas in the future and whether the solution has long-term viability. That's because the assumption that the captured CO_2 can be stored indefinitely is optimistic and unproven.

Moreover, producing natural gas releases methane, a potent greenhouse gas that, in 20 years, can warm the air 86 times more than CO_2. Also, inevitable methane leaks happen during drilling, extraction, and transportation.

According to the Global Hydrogen Review 2021, released by the IEA, the demand for hydrogen stood at 90 MT in 2020, produced almost exclusively from fossil fuels, generating 900 MT of CO2 emissions.

Considering both the uncaptured CO2 and the large emissions of unburned fugitive methane emissions, scientists claim that to create blue hydrogen, the carbon footprint is more than 20 per cent higher than using either natural gas or coal directly for heat, or about 60 per cent more than using diesel oil for heat.

Therefore, to say that hydrogen presently produced is a low-emission or zero-emission fuel is far from the truth.

The IEA has urged countries to take rapid and more decisive steps to lower the barriers holding green hydrogen back from faster growth. It will be important if the world has to reach the net-zero emission target by 2050.

The UAE is taking great strides in developing a robust green hydrogen industry. The country has the natural resources, technological expertise, and political will to become a key player in the global green hydrogen supply chain.

Though the technology is still nascent, experts claim that until commercial-scale green hydrogen production is successfully deployed and becomes cost-competitive, blue hydrogen offers a bridging technology that can be used to support the transition to a decarbonised energy system.

However, that denotes a 'halfway solution' and there's no time to lose. The climate crisis is driving people out of their homes in vulnerable countries and causing massive losses to the Earth's biodiversity.

The frequency, intensity, and duration of natural disasters that destroy people's lives and livelihoods are aggravated yearly.

The climate crisis is amplifying displacement and making life harder for those already forced to flee due to made-made conflicts and territorial wars, says a report by the United Nations High Commissioner for Refugees (UNHCR).

The organisation says that climate change and displacement are increasingly interconnected.

As extreme weather events and environmental conditions worsen with global heating, they contribute to multiple and overlapping crises, threatening human rights, increasing poverty and loss of livelihoods, straining peaceful relations between communities, and ultimately, creating conditions for forced displacement. The conflict for food and clean water will escalate if a check is not kept on the climate crisis.

Meeting climate pledges requires faster and more decisive action. Unless we race against time, future generations will face the consequences and pay for our slackness.

6

Confusing Signals

I would rather be of a clear mind and decision with the wrong club than with an unclear mind and the right club.

Walter Hagen,
Sportsperson

In prelude to the Paris Agreement, the fifth assessment report of the IPCC 2014, delineated the permissible amount of CO_2 emissions compatible with limiting global warming to specified levels.

It urged nations to 'achieve a balance between anthropogenic emissions by sources and removals by sinks of GHGs in the second half of this century', but without specifying the number of CO2 equivalents to be removed.

It did not use the terms 'net-zero', 'carbon neutrality', and 'climate neutrality', employed interchangeably in the climate discourse. While these terms share a common goal of reducing GHG emissions, they bear nuanced distinctions in meaning and scope.

Net zero encompasses the attainment of equilibrium between GHG emissions generated and those removed from the atmosphere, including CO2. It mandates exhaustive emissions reduction at source, with the extraction of any natural or technological residual emissions. The approach aligns with the Paris Agreement, rendering it a compelling goal for corporate climate endeavours.

Net-zero amalgamates an array of commitments and initiatives geared towards 2050. It hinges on sustaining a balance between GHG emissions and their removal from the atmosphere via natural or technological means, such as afforestation, reforestation, or carbon capture and storage.

In contrast, carbon neutrality centres on offsetting CO2 emissions produced by an entity, product, or service by securing equivalent CO2 reductions or removals. It addresses CO2 emissions, omits other GHGs, and does not require emissions reduction at their source. Carbon neutrality can be accomplished through investments in carbon sinks such as forests, oceans, or carbon credits, including renewable energy projects or emission reduction programmes.

Businesses often embrace this approach to display their environmental commitment, although it may fall short of limiting global warming to 2°C.

Similarly, climate neutrality parallels carbon neutrality but extends its purview to cover all types of GHGs, not just CO2. Achieving climate neutrality signifies that an entity exerts no net influence on the climate system by diminishing emissions to zero or offsetting them through removals or reductions. Low-carbon technologies, like renewable energy and electric vehicles, are paths to realising climate neutrality.

Global leaders and groups invoke this concept to exemplify their dedication and leadership in combating climate change.

Net-Zero In Spotlight

Before the COP26 held in 2021 in Glasgow, UK, the UN orchestrated the 'race for a net zero' campaign, soliciting support and leadership from diverse sectors, including businesses, cities, regions, investors, and educational institutions.

Over 120 countries voiced their intent to attain net-zero emissions by 2050, with over 3,000 non-state actors, constituting over 15 per cent of the global economy, joining the campaign. Still, as of 2024, a huge gap persists between commitments and actions.

Despite the potential advantages of net-zero policies leading to improving public health, bolstering energy security, and creating novel economic prospects, these fell short.

Net zero is built upon the premise of balancing emissions by removing an equivalent quantity from the atmosphere. Unlike the Kyoto Protocol's concept of carbon neutrality, which introduced binding emissions reduction targets and mechanisms like emission trading and carbon offsetting for developed nations, net-zero lacks binding measures.

What is the Kyoto Protocol?

Adopted on December 11, 1997, and applied on February 16, 2005, with 192 nations, the Kyoto Protocol operationalises the United Nations Framework Convention on Climate Change (UNFCC) by committing industrialised countries and economies in transition to limit and reduce GHG emissions according to agreed individual targets.

It is based on principles and provisions of the Convention and follows its annex-based structure. It binds developed nations and places a heavier burden on them of 'common but differentiated responsibility and respective capabilities', recognising that they're responsible for high-level GHG emissions.

Numerous nations have announced intentions but failed to delineate comprehensive implementation plans and enacted laws mandating net-zero emissions without specifying the execution strategy. Therefore, the path to net zero remains murky, necessitating nations to adjust their targets.

Net-zero policies harbour multiple shortcomings. It demands an intricate, unpredictable, interdependent, and transformative shift that may remain elusive without a coordinated transition plan, underpinned by research, investments, policies, and regulations.

The repercussions may be asymmetric, confronting technical, economic, and social hurdles. Achieving it requires widespread deployment of low-carbon technologies, all of which contend with obstacles like mineral scarcity and exorbitant costs.

It also mandates substantial changes in governance at every step – an ardous feat within the constrained timeframe, marked by resistance from deep-rooted interests, lethargy, and lack of awareness. Net-zero policies are not a solution for addressing the climate crisis; they must complement other policies and measure its root causes and consequences.

Aspiration To Reality

In 2021, the Massachusetts Institute of Technology (MIT) Global Change Forum raised carbon budget matters, decarbonising energy and industry, nature-based solutions, climate and health, negative emission technologies, and policy design.

The forum underscored the urgency of interactions between human and natural systems, emphasising collaboration and innovation across all stakeholders, involving governments and international institutions. The experts called for unprecedented and immediate action from all to transform net zero from aspiration to reality.

However, the fact is that the Earth has already warmed by about 1°C compared to the pre-industrial era. Also, the current net-zero policies are inadequate to prevent further warming due to the lack of ample measures and actions. The future emissions are uncertain and unreliable because the atmosphere has been affected by past emissions for a long time.

Where Did We Go Wrong?

Things should have been done differently to give us a climate-safe environment. The stipulations in the deal are not strict enough to prevent or delay the progression of climate change. Also, lack of enforcement undermines its effectiveness considerably. The prime example is the US, the second-largest global emitter, which walked out of the Paris Agreement without being reprimanded. Action against it would have set an example for others.

Donald Trump's U-Turns

In 2017, US President Donald Trump announced his intention to withdraw from the Paris Agreement, formally notifying the UN. It is to be noted that since the industrial era began in the mid-1800s, the US has emitted more cumulative CO2 into the atmosphere than any other country.

The emissions fell too slowly to avoid catastrophic warming because the Trump administration rolled back carbon pollution limits from power plants, cars, trucks, and fossil fuel operations.

Under the Paris Agreement, the country promised to reduce its emissions by about 25 per cent by 2025. However, observers felt that Washington weakened the agreement and then walked away. They alleged that the US did it first with the Kyoto Protocol and later with the Paris Agreement.

The US walked away from the Kyoto Protocol with accusations that China and India were not reducing emissions, so it would not participate in a 'flawed' protocol. However, to 'facilitate' the US participation in the Paris Agreement, all 194 countries pledged emissions reduction, even though the majority did not contribute to climate change.

Trump dismantled the 'modest' policies and initiatives of former US President, Barack Obama to reduce fossil fuel consumption and GHG emissions.

He refused to fund the Global Climate Change Initiative (GCCI), including the Green Climate Fund (GCF), set up to help developing nations adapt to climate change and move to low-carbon technologies. This was the key attribute of the Paris Agreement by the wealthy countries like the US to provide financial support to developing nations.

It dealt a death blow, weakening the character of the Agreement, which was seen as meant for the US, by the US.

However, after a few months in office, US President Joe Biden re-entered his country into the Agreement.

The experts denounced that with 5 per cent of the world population, the US should not have been allowed to gamble with the lives and livelihoods of the rest of the world.

Seeing the current results, it's time the nations came together to see how the Agreement could be amended to make it effective and ambitious. Countries ratifying it knows they can get away with it and continue polluting and emitting GHGs at an alarming rate.

Besides, the Paris Agreement has overfocused on CO_2 rather than concentrating on all GHGs. To reach carbon neutrality by 2050, it overlooked the more potent gases, such as hydrofluorocarbons (HFCs), some of which are 4,000 times more potent as GHGs than CO_2. Their use is increasing by 10 per cent annually.

Ever since the Paris Agreement, developing countries at the forefront of experiencing devastating climate disasters have demanded that developed nations – the largest emitters of greenhouse gases – compensate them for the harms faced due to climate change.

As mentioned earlier, though Africa contributed the least to climate change, it has always been castigated. The imbalance in accountability must not be allowed and funding the affected nations is a crucial starting point to rectify the prolonged systematic injustice to vulnerable communities.

Methane – A Spoil Sport

It is noted that apart from CO2, methane (CH4) is another main player destroying the planet. Even though China, India, and Russia are primary contributors of CH4 into the atmosphere, of the 119 countries who signed a pledge at COP26 in Glasgow to reduce CH4 output, only 40 (meaning one-third), actively introduced legislation and made commitments.

Predominantly, crops, livestock, and leaking gases from landfills produce CH4.

India Speaks

However, India, under the Indian Council of Agriculture Research (ICAR), aided by other governmental organisations, has taken measures to develop technologies to cut down CH4 emitting in the atmosphere.

Farmers producing rice and paddy – the main source of consumption in the country and which produce CH4 – are encouraged to diversify their crops to millets, pulses, oilseeds, maize, cotton, and agroforestry.

In 2022, India upped its targets of reducing the energy intensity of growth and share of installed capacity of non-fossil fuel power by 2030. The country is well on track to meet these goals. While the state governments are investing in public transportation in major cities, the public is contributing with their lifestyle changes to lower GHG emissions.

Therefore, asking all developing economies to shoulder the burden of keeping GHG emissions in check is incorrect. The onus of action also rests on advanced geographies.

Bappaditya Mukhopadhyay, a professor of analytics and finance at the Centre of Excellence for Sustainable Development, Great Lakes Institute of Management, Gurugram, India, has implored that to honour the commitment to reduce GHG emissions, "A mechanism should be devised that calculates before the beginning of every year the GHG reduction each country must necessarily achieve."

"A monetary equivalent of the same should be paid by respective countries into a global pool. Then, depending on the actual GHG emissions throughout the year, the balance would be credited back to the country from the global pool.

The commitment amounts for every country can be drawn from historical and projected GHG emission levels," he added.

The suggestion seems logical if countries consider a clean environment a basic human right.

7

The Wisdom Bridge

Climate change does not respect borders; it does not respect who you are – rich, poor, small, and big.

Ban Ki-moon,
Former Secretary-General of the United Nations

The Middle East and North Africa (MENA) region is an interesting case study for global energy transformation. One of the most dependent on fossil fuels globally for its economic growth, after oil and gas were discovered a century ago, it commercialised to stoke up the booming industries in Europe and the US.

The MENA region contains nearly 60 per cent of the

world's oil reserves and 45 per cent of natural gas reserves.

However, it faces several challenges and risks related to its reliance on fossil fuels due to price volatility, environmental degradation, and geopolitical conflicts. Therefore, countries in the region are transitioning to a post-fossil fuel era by diversifying their economies and investing in renewable energy sources that would lead to innovation and a balanced atmosphere.

The IRENA report states that the MENA region has a huge potential for renewable energy development, especially solar and wind power. It added that by 2030, it could generate more than 60 per cent of its total electricity from renewable sources.

UAE: How It Should Be Done

The UAE has been a leader in modernisation and sustainability, with a vision of becoming a global hub for a green economy and clean energy. It hopes to increase its share of clean energy to 50 per cent by 2050.

This Middle Eastern country has invested in various renewable energy projects, including the Shams 1 solar thermal plant, one of the largest in the world, and the Mohammed bin Rashid Al Maktoum Solar Park, planned to be the largest single-site solar park globally by 2030. The UAE is also home to the well-planned Masdar City, which aims to be one of the most sustainable cities in the world.

According to BP statistical reviews, it is also one of the 10 largest oil-producing countries with proven reserves of 100 billion barrels.

Judicious Investment In Research

The Emirates is at the forefront of R&D, driving economic diversification and addressing local and global challenges. Through strategic investment, it creates new technologies, fosters sustainable practices, leverages existing universities and institutions' expertise, and solidifies its reputation as a knowledge-based economy.

During the 2024 Holy Month of Ramadan, prominent professors and researchers shared their insights on advancing papers to tackle urgent sustainability and climate change matters. The dialogue explored certain challenges and reviewed the progress of scientific studies.

These included aligning research with societal needs, funding aspects, knowledge transfer, transforming the research environment, global collaboration, implementation, and involving the youth as agents of change.

In the current scenario, such activities are crucial in supporting innovation and sustainable practices that drive the development of technologies and policies essential for promoting sustainability. Research also aids in understanding and addressing environmental challenges and training for future complications.

Just as universities, research institutions, and innovation parks form the backbone of the R&D initiatives of the federation of seven emirates, many distinguished researchers, professors, scientists, and laboratory technicians are a driving force behind optimising high-quality results. Together, they contribute to building a promising future and engage in challenging and intriguing topics.

The country's commitment to research and innovation is not just about advancing self-interests; it contributes to the global pursuit of knowledge and sustainable development that spills beyond its borders. By sharing its findings and collaborating with international partners, the Emirates is helping to shape a better future for all.

Leadership Vision

One of the essential elements that can make a significant mark in achieving sustainability goals is the country's leadership. The UAE President, His Highness Sheikh Mohamed bin Zayed Al Nahyan, is proactive on the development agenda and regulates climate change policies dynamically.

He declared in 2015, "In 50 years, we will celebrate the last barrel of oil." He knew early on to move on the right track. In fact, one of his first decisions after assuming the top position in 2022 was the pledge to invest an additional $50 billion to scale up climate action by deploying clean energy solutions locally and globally.

Markers For Prosperity

In the 2030 agenda, 17 Sustainable Development Goals (SDGs) are seen as a roadmap for economic growth, social inclusion, and environmental protection. The UAE has shown steadfast commitment by devising a robust plan to chart a green ecological future for all.

Blessed with the world's seventh largest oil and natural gas reserves, even then, it has taken vigorous steps to diversify its energy in tune with advancement.

Over 70 per cent of its economy is non-oil based, built on a lengthy record of accomplishment spanning three decades of environmental stewardship.

Even though it contributes less than 0.6 per cent of global emissions, the Emirates is solemn as a climate action leader, embarking on a journey to reduce emissions globally.

In line with SDG Goal 13: Climate action – The UAE was the first in the MENA region to commit to reaching net zero by mid-century. It plans to achieve the target by driving decarbonisation and integrating workable energy across all sectors.

The nation is rapidly expanding its clean energy capacity, operating three nuclear power reactors that provide energy to its grid. It houses three of the largest and low-cost solar plants in the world.

Its green hydrogen project, implemented by the Dubai Electricity and Water Authority (DEWA), in collaboration with Expo 2020 Dubai and Siemens

Energy at the Mohammed bin Rashid Al Maktoum Solar Park, is the first of its kind in the MENA region to produce hydrogen using solar power.

To accommodate future utilisation in energy generation and transportation, the pilot project tests different hydrogen uses.
Even in the swiftest energy transition scenario, the world will require oil and gas; thus, by geology and design, the country's hydrocarbons are among the least carbon-intensive in the world. Yet, it is meticulous in lowering carbon impact and is the first in the region to deploy industrial-scale carbon capture, utilisation, and storage technology.

In support of SDG 14: Life below water and SDG 15: Life on land – The UAE is set to meet its goal of planting 100 million mangroves by 2030, as these provide natural infrastructure and protection to populated areas by absorbing storm surges during cyclones and hurricanes.

Reflecting on SDG 17: Partnerships for the goals – A key enabler of UAE's holistic approach to building a better future is to form effective partnerships that support its aims. It joined forces with the US in launching the Agriculture Innovation Mission (AIM) for Climate initiative at COP26.

The game-changing drive sought to increase investment to support and fund climate-smart agriculture and food systems innovation to address food insecurity and global warming. Recognising that these contribute one-third of global human-caused

GHGs, by 2022, the 300 government and non-government collaborates increased investments by more than $8 billion.

AT COP27, the UAE-entered partnerships took two extraordinary leads, expecting far-reaching climate benefits.

One was the US-UAE Partnership for Accelerating Clean Energy (PACE), with an ambitious goal of catalysing $100 billion in financing and investment and deploying 100 GW of clean energy globally by 2035 to advance the transition.

The second was with Indonesia, for the Mangrove Alliance for Climate (MAC), which will fast-track the conservation and restoration of mangrove ecosystems as an effective nature-based solution.

It hosted COP28 with the same belief that partnerships are the best tools to slow down climate change. It strives to work closely with all participants to ensure they can contribute to bold, ambitious, and practical solutions while prioritising inclusivity, enhanced accountability, and transparency.

The nation's leadership hopes that future generations can grow and prosper in a sustainable world.

The countries that have made significant progress in the MENA region:

- **Morocco:** The North African country, bordering the Atlantic Ocean and the Mediterranean Sea, has been a pioneer in renewable energy development, targeting to achieve 52 per cent of its electricity from renewable sources by 2030. Morocco has invested heavily in solar power projects, including the Noor Ouarzazate Complex, the world's largest concentrated solar power plant.

- **Egypt:** Like its regional counterparts, the country linking northeast Africa with the Middle East has also been pursuing an ambitious renewable energy strategy to reach 42 per cent of its electricity from renewable sources by 2035. Egypt has potential solar and wind power, especially in the Sinai Peninsula and the Red Sea coasts. It has launched several large-scale projects, including the Benban Solar Park, which on completion is likely to be the largest solar photovoltaic plant in the world, and the Zafarana Wind Farm, one of the largest in
  Africa.

- **Saudi Arabia:** Though Saudi Arabia generates most of its power through fossil fuels, the Kingdom has an ambitious programme for a clean energy revolution, whereby the government aims to invest a total of $101 billion to increase renewable energy to 50 per cent of power generating capacity of about 58.7 GW by 2030. It plans to plant 10 billion trees to cut 278 million tonnes of CO2 by the decade's end and hit net zero by 2060.

- **Oman:** The nation has committed to net-zero emissions by 2050 by expanding its electricity-generation capacities through renewable power projects, deriving at least 30 per cent of electricity from renewables by 2030 on wind and solar projects.

- **Qatar:** By producing more than 2,000 kilowatt hours of solar power per square metre annually, the nation has chalked out the Qatar National Vision 2030. It aims to expand the renewable power generation capacity to approximately 4 GW by 2030 with distributed solar generation contributing around 200 MW. This will help to decentralise power generation and ease the burden on the national grid to supply power.

However, many challenges and barriers must be overcome, such as policy and regulatory frameworks, financing and investment mechanisms, public awareness and participation, regional cooperation, and integration with Artificial Intelligence (AI).

Incorporating AI in climate change research simplifies the analysis of the intricate dynamics of the systems, enabling the development of strong strategies and policies to counteract and mitigate the negative consequences.

Reality of AI
It has become an invaluable tool in predicting trends and patterns, aiding in effective

environmental policies, and enabling accurate forecasts.

Instrumental in optimising energy consumption and smarter resource utilisation, it supports sustainable practices through automation and enhanced performance. Due to the ease of data analysis, the vast information available can provide accurate solutions for sustainable development.

Technologies are created for environmentally friendly conditions and socially equitable and economically viable solutions. In climate change, sustainable AI will optimise energy in buildings and reduce GHG emissions.

AI enhances understanding of climate change's

impacts by utilising its capacity to process and analyse large datasets. The algorithms monitor shifts in environmental elements such as icebergs and forest cover.

Furthermore, it evaluates and forecasts the potential climate change risks, essential for effective management and preparedness.

A prime example is artificial neural networks for predicting weather patterns, including the likelihood of storms and hurricanes.

A United Nations report highlights how AI helps in combating climate change:

Weather: AI assists in processing volumes of data to extract knowledge, improve predictive models, and help the vulnerable in Burundi, Chad, and Sudan to investigate past environmental changes.

For example, the MyAnga app helps Kenyan pastoralists brace for droughts. With data from global meteorological stations and satellites sent to their mobile phones, herders can better manage their livestock and save hours of scouting for green pastures.

Disaster prevention: Helping communities to prepare for climate disasters, AI-driven initiatives target high-risk areas and feed into local and national response plans. For areas susceptible to landslides, mapping can help local groups plan and implement sustainable development measures, reduce risks, and ensure the safety of residents in unguarded places.

'The Early Warnings of All' initiative ensures everyone on the Earth is protected from hazardous weather, water, or climate events through warning systems by the end of 2027.

Tracking pollution: Using AI, susceptibility maps can support local governments in making decisions to improve public health and urban resilience.

Additionally, AI can improve urban planning, traffic, and waste management, making cities more sustainable and liveable.

Carbon neutrality: A critical catalyst in realising global carbon neutrality goals, AI algorithms can minimise environmental impact and maximise efficiency. To realise the global goal for affordable and clean energy for all by 2030 (SDG 7), AI can optimise grids and increase the efficiency of renewable sources. Predictive maintenance using AI can also reduce downtime in energy production to diminish carbon footprint.

Fast fashion: An industry with a record of high emissions, fashion can benefit from AI-driven R&D to accelerate innovation. The $2.4 trillion global industry that employs approximately 300 million people across the value chain, many of whom are women, is expected to grow in the coming years. AI can optimise supply chains to reduce waste, monitor resource consumption, and promote sustainable manufacturing processes.

Fast food: Agriculture is a high-emission sector, accounting for 22 per cent of global GHG emissions. But AI can bring a change. With farmers and corporations facing extreme weather, water scarcity, and land degradation, AI can help optimise their practices, reduce waste, and minimise the environmental impact of food production.

It can balance supply and demand, facilitate the integration of renewables into energy systems, and reduce the reliance on fossil fuels. It is pivotal in building climate-resilient agri-food systems that are efficient and adaptable to climate challenges.

Robin Bordoli, the CEO at Weights & Biases, the AI developer platform, sums it up by stating, "What makes AI different from other technologies is that it will bring humans and machines closer. AI is sometimes incorrectly framed as machines replacing humans. It's not about machines replacing humans, but machines augmenting humans."

8

War Proliferates Global Carbon Emission

Reversing global warming will take a World War II level of mobilisation. It is the work of tens of millions, not hundreds of thousands.

Van Jones,
Political Analyst, and civil rights advocate

War is never a mandate for rational reasoning. The ongoing conflict between Israel and Hamas has displaced millions and left thousands killed, thus dividing the world into groups. It has affected the climate with the heat trapped in the atmosphere leading to warmer regions.

In a report published by the Social Science Research Network, the global warming emissions generated during the first two months of the war in Gaza, since late 2023, were higher than the annual carbon footprints of more than 20 of the world's most climate-vulnerable nations.

Benjamin Neimark, a senior lecturer at Queen Mary, University of London, and co-author of the research stated, "This study is only a snapshot of the larger military boot print of war... a partial picture of the massive carbon emissions and wider toxic pollutants that will remain long after the fighting is over."

One cannot take an eye off the Russia-Ukraine conflict that has entered its third year, after beginning in February 2022. The global estimates are mind-blowing, but the UN data states that constant bombardment by Russia in Ukraine has led to blanket contamination of air, water, and soil due to chemicals and toxic gases.

Nearly 30 per cent of Ukraine's arid land is not fit for plantation. Moreover, wildfires and demolished buildings due to air and ground strikes have led to rising pollutants: multiplying respiratory diseases.

Wars are fought to change the demographic boundaries of nations as per one's strength and sustainability.

Going by numbers, more than 230 million people died because of wars from 1900 until date. The most talked about and read conflicts of the last century were the two World Wars, the Korean War, the Vietnam War, the Syrian Civil War, and the present Gaza War. The global conflicts led to about 240,000 people being killed in 2022 alone.

GHG emissions are directly related to conflicts. The various pollutant gases, CO2, methane, nitrous oxide, and fluorinated gases have enveloped the Earth's atmosphere, resulting in people gasping for air.

A study by the IEA and the Stockholm International Peace Research Institute, says that the world's armed forces emitted about 1.2 billion tonnes of CO2 in 2019 - about 2.3 per cent of the global total carbon emissions.

The US, the largest contributor to military operations, emitted about 205 million tonnes of CO2e (carbon dioxide equivalent) in 2019, corresponding to 0.4 per cent of the global total. However, these estimates do not account for the emissions from industries that supply arms and equipment, nor the emissions from the environmental damage and displacement caused by war and conflict.

In the March-April 2024 scenario, the amount of GHG in Gaza was estimated to be 70 million tonnes. Though numbers may vary, some estimates use a simple method of multiplying the number of casualties by the average carbon footprint of a person. The method considers only the reduction in emissions - 1.2 million tonnes of CO2e annually.

However, it's assumed that the reduction is taken into account from emissions from military strikes, infrastructure damage, and reconstruction, thus negating the larger picture of emissions due to civilian deaths.

Unfortunately, the GHG emissions from wars and military operations are not well reported, accounted for, or documented by many countries. This is due to ambiguity in international climate agreements, which do not require or oblige nations to include military emissions in their reports.

This excuse occurred in 1997, during the Kyoto Protocol negotiations, when the US insisted on excluding emissions from multilateral operations, activities that involve more than one country, and ships and aircraft associated with global transport.

The Paris Agreement does not include exemptions for military emissions; however, it leaves it to the discretion of the nations to declare it.

Therefore, the available data relies on estimates from independent researchers and organisations that vary depending on the source and method used in documenting the GHG impact.

During COP27 in Glasgow, in 2021, climate activists, academics, and civil society organisations stressed the inclusion of military emissions on the formal agenda of the UN meeting. More than 200 civil society groups, including Amnesty International and Human Rights Watch, signed the Conflict and Environment Observatory's call for governments to reduce emissions from military operations.

The groups argued that war and climate change were interlinked and that addressing one required addressing the other. They also urged the international community to adopt a human rights-based approach to climate action and ensure that the most vulnerable and affected people were protected and looked after.

However, with more than 2 per cent of the global emissions not reported and remaining data classified, global warming remains under limited control.

War Not A Solution

Fortunately, Gen Z is active and vocal in their views on the impact of war on civilians and climate change. Peace coalitions like the Global Coalition on Youth, Peace and Security (GCYPS) and the United Network of Young (UNOY) Peacebuilders are asserting themselves to strengthen youth participation in bringing harmony to the negotiating table of the warring factions.

Individuals are also trying to prevent and resolve conflicts by taking the initiative to reach the masses through essays, climate competitions, and placard marches within their vicinity.

9

Are We Missing The Message?

What is the use of a house if you haven't got a planet to put it on?

Henry David Thoreau,
American Philosopher

Former US Assistant Attorney General William Ruckelshaus said, "Nature provides a free lunch, but only if we control our appetites." The first Environmental Protection Agency (EPA) administrator meant that we must give back what we take and take only as much as we need.

Sadly, as is the human tendency, greed took precedence. The result is the calamities all around. Ignoring the warnings and calls for taking stock of the dwindling quality of life on the planet, even as we missed umpteen messages, there was also the significant nationally determined contributions (NDCs) goal set by the think tanks, which did not provide well-defined guidelines.

The 2015 Paris Agreement stated that global warming should be below 2°C or preferably 1.5°C. However, the NDCs failed to communicate how much carbon emissions needed to be reduced. It would be good if policymakers meditated on this crucial point at the COP29 annual UN climate meeting in Azerbaijan in November.

Until then, we need to work on a war footing to eliminate the dreadful impacts, underscoring the urgency of addressing climate change and its effect on our planet and habitat.

The key points to keep in mind regarding the aftermath of burning fossil fuels and introducing new gases in the atmosphere include:

Impact On The Earth
- *The global surface temperature will increase and lead to widespread habitat loss on land and sea.*
- *Wars and calamities including heatwaves, droughts, storms, and heavy precipitation will displace all living beings.*

Impact On Biodiversity
- *The IPCC reports indicate that up to 30 per cent of species could die if global mean temperatures rose by 1.5 to 2.5°C.*

- *Rapid glacial melt influences ocean currents, as the massive amount of freezing water enters warmer oceans slowing the currents. As the ice on land melts, sea levels will continue rising and affecting numerous species.*

Impact On Coastal Areas

- *Thermal expansion of oceans and polar ice melting will cause sea levels to rise, threatening more than 90 per cent of the Costa regions during the 21st Century.*
- *The increased serenity of land will lead to habitat changes and cause loss of livelihood.*
- *Since more than 90 per cent of the world's infrastructure is in coastal areas, it will affect buildings and structures, contaminate drinking water, and interrupt emergency services.*

Impact On Ecosystem Services

- *The decline in agricultural aspects, infrastructure, and tourism will cause many economies to fail.*
- *Changes in nutrient cycling, pollination, decomposition, and water purification.*

Without numerous changes, life on Earth would be impossible to sustain.

However, as they say, hope is being able to see that there's light despite the darkness, the youth are the torchbearers of the infinite tools that lay before them. With robust technology at their fingertips, present-day activism is not nebulous, as the students discuss the perils of environmental degradation in schools, colleges, and public forums.

It's time to live up to the words of Martin Luther King, Jr, "We must accept finite disappointment, but never lose infinite hope."

10

Time For The Youth To Reset The Button

Trust me, I live climate change, my friends and family live it too.

Nikosilathi Nyanti,
UNICEF activist, Zimbabwe

The saying, 'Nothing brightens the life of any youth better than knowledge and ability to affect a positive change,' fits in the thick of things the youngsters are in, as they are the ones to suffer the most on inheriting a smoky planet from elders.

All eyes are on them to turn around the Earth's climate with support from seniors as warmer temperatures have altered the living pattern, leading to an abundance of allergens causing lung infection, rising asthma cases, and cardiac arrests at a young age.

The above-normal temperatures have caused early flower blooms, causing a mismatch between pollinators and plants.

However, the youths are not passive victims but active agents of climate change due to the robust use of electric appliances, gadgets, processed foods, and pollution-emitting transports.

Good To Go!

The comforting news is that the younger generation is leading the global movement to demand more actions from their governments and other entities to tackle climate change. Utilising their skills and creativity to raise awareness, educate others, and implement solutions, they strive for a better present and posterity.

Fortunately, they have better tools and opportunities to make a difference. With access to education, technology, and science, they can alert and address the challenges of climate change from any corner of the world.

Utilising the space in the international arena to hear their pleas, the young influence decision-making processes that affect their future.

For example, at the COP forums, they organise and participate in events and activities including the Conference of Youth (COY), an annual event that prepares youngsters to participate in debates.

The Youth and Future Generations Day emphasises

initiatives and perspectives on climate change. At the COP meetings, they have engaged with different stakeholders, such as governments, civil society, media, and businesses, to speak out and influence opinions. The youth are expected to advocate ambitious and inclusive climate actions soon.

The UN designated August 12, as International Youth Day (IYD) in 2000. It was conceptualised by young people at the World Youth Forum in Vienna, Austria, in 1991. Hosted by the Government of Portugal alongside the UN in 1998, it was adopted in a resolution by the World Conference of Ministers Responsible for Youth.

With education the key to their agenda, the purpose is to draw attention to cultural and legal issues and encourage participation, especially of those, affected by poverty.

The theme is different every year, reflecting the current challenges and opportunities. The subject for 2023 was 'Green Skills for the Youth: Towards a Sustainable World'.

It highlighted the need to develop and enhance their knowledge, abilities, values, and attitudes that enable them to support and nurture a sustainable and resource-efficient society. It comprised technical skills to facilitate green technologies and processes, and transversal abilities that foster environmentally sustainable decisions in life and at work.

The theme also recognises the importance of youth as change champions, entrepreneurs, and innovators who can contribute to the green transition and the achievements of the SDGs.

It's My World

As per the United Nations Children's Fund (UNICEF) US report, all over the world, young people are aware of climate change. UNICEF launched 'The Green Rising' project with an ambitious plan to mobilise 10 million young people to take concrete climate actions in their countries and communities by 2025.

"We need young people in the rooms where decisions about our planet are being made because young people are not only on the front line of the climate crisis, but we are also on the front lines of the fight for climate justice," said the organisation's Goodwill Ambassador Vanessa Nakate.

UNICEF *feels young people are eager to:*

- Lead projects that build climate resilience at local schools, hospitals, community centres, and neighbourhoods.
- Get trained in green skills and drive the green transition across sectors.
- Protect ecosystems by preserving water sources and forests, using cleaner water, and turning waste into energy.

- Work in local communities to help adapt to future climate events – using gender-transformative approaches.
- Adopt sustainable lifestyles and advocate offline and online for others to do the same.
- Encourage governments and corporates to enact policies, adopt sustainable practices, and fund climate adaptation measures.

Youth climate actions in a nutshell:

- *In Burkina Faso, kids are learning to install solar power*
- *In Lebanon, they are setting up systems to turn biowaste into fuel.*
- *In Cambodia and Kazakhstan, they are aiming for single-use plastics.*

- *In Burundi, they build clean energy stoves, sans wood for cooking.*
- *In India, they are trained in water conservation in parched states.*
- *In Kyrgyzstan, they focus on disaster management during weather crises.*
- *In Malaysia, they impart indigenous knowledge to protect the environment.*
- *In Maldives, the 'Muhyiddin Scouts' address coral bleaching.*
- *In Mongolia, they monitor air pollution levels touching red zones.*

UNICEF has long prioritised youth engagement around climate issues, providing platforms to share their ideas and advocate for global actions.

In 1995, Michael Jackson, the 'King of Pop', regarded as one of the most significant cultural figures of the 20th century, wrote and sang the 'Earth Song', creating environment awareness due to conflicts and climate change. Pressing the right buttons, the lyrics have interwoven questions for us to answer:

What about sunrise? What about rain?
What about all the things that you said we were to gain?

Acknowledgments

This book would not have been possible without the support of my family, which means everything to me. I owe them the greatest debt for making it possible - first in living it and then writing it.

I am especially grateful to my caring daughter Mais for her unending inputs on the perils of climate change and how the youth can be moulded by making the subject human-centric. She suggested that in my speeches on varied platforms, I must include how the changing climate might limit our choices of favourite sandwiches or preferred beverages due to environmental severity.

My sincere acknowledgments are owed to friends and well-wishers for being generous with their time and providing invaluable insight into the numerous articles I wrote for newspapers and periodicals.

I am highly obliged to my friend Chauki Rafeh for his encouragement and feedback on my book-writing journey and for believing it would take shape one day.

I am grateful to Lovely Reanolyn for meticulously verifying facts and figures and making them informative and educative.

My special thanks to **Nilima Pathak** for her research and assistance and for piecing together my thoughts with sensitivity and professionalism.

About the Author

Dr. Abdulla Belhaif Alnuaimi is a senior fellow at the American University of Sharjah and an honorary professor at Heriot-Watt University, Dubai.

His distinguished career spans education, politics, and engineering.

As the UAE's Minister of Climate Change and Environment (2020-2021), he established laboratories and centers for research and applied studies in the region. In addition to his ministerial responsibilities, he headed the UAE Council for Climate Change and Environment, led the National Biosecurity Committee, and served as Chairman of the Circular Economy Council.

His political career includes serving as Minister of Infrastructure Development (2013-2020). He was also Chairman of the Sheikh Zayed Housing Programme and Chairman of the Federal Transport Authority– Land & Maritime (2017-2019).

Dr. Abdulla has been Chairman of the Sharjah Consultative Council (2023-Present) and served as Undersecretary in the Ministry of Infrastructure Development, formerly the Ministry of Public Works (2003-2013).

In his professional career, he worked as an Engineering Manager at the Department of Civil Aviation in Abu Dhabi (1993-2003) and as Director of the Water Distribution Department at the Ministry of Water and Electricity (1981-1993).

He is a recipient of several awards and recognitions, including the Who Is Who International Awards in Greece (2024), where he was honored as the World's Eminent Man in Education & Sustainability.

He also won the Kanz Al Jeel Awards from the Department of Culture and Tourism, Abu Dhabi (2023), a Fellowship Certificate from the London Institute of Civil Engineers, United Kingdom (2019), and the UAE's Global Maritime Community Award at the 15th Sea Trade Maritime Awards for the Middle East, Indian Subcontinent, and Africa (2018).

He holds a PhD in Engineering Project Management from the University of Reading, United Kingdom (1990), and a Bachelor's degree in Mechanical Engineering from the University of Wisconsin, United States (1980).

A keen poet, Dr. Abdulla has written seven poetry books, including *'A Love Renewed,'* a collection that maintains the authenticity of Nabati poetry, showcasing rich poetic vocabulary and imagery

"ظهر الفساد في البر والبحر بما كسبت أيدي الناس ليذيقهم بعض الذي عملوا لعلهم يرجعون "
صدق الله العظيم

"Corruption has appeared throughout the land and sea by [reason of] what the hands of people have earned so He may let them taste part of [the consequence of] what they have done that perhaps they will return "

الآية 41. من سورة الروم

www.ingramcontent.com/pod-product-compliance
Lightning Source LLC
Chambersburg PA
CBHW022047210326
41519CB00055B/1097